John Tarilanyo Afa

Overview on the Problems of Rural Electrification in Bayelsa State

GRIN Verlag

Bibliografische Information der Deutschen Nationalbibliothek:

Die Deutsche Bibliothek verzeichnet diese Publikation in der Deutschen National-
bibliografie; detaillierte bibliografische Daten sind im Internet über http://dnb.d-
nb.de/ abrufbar.

Imprint:

Copyright © 2011 GRIN Verlag GmbH
Druck und Bindung: Books on Demand GmbH, Norderstedt Germany
ISBN: 978-3-656-41277-9

GRIN - Your knowledge has value

Der GRIN Verlag publiziert seit 1998 wissenschaftliche Arbeiten von Studenten, Hochschullehrern und anderen Akademikern als eBook und gedrucktes Buch. Die Verlagswebsite www.grin.com ist die ideale Plattform zur Veröffentlichung von Hausarbeiten, Abschlussarbeiten, wissenschaftlichen Aufsätzen, Dissertationen und Fachbüchern.

Visit us on the internet:

http://www.grin.com/

http://www.facebook.com/grincom

http://www.twitter.com/grin_com

JOHN TARILANYO AFA

CURRICULUM DESIGN

TECHNICAL REPORT

PRE 677:

OVERVIEW ON THE PROBLEMS OF RURAL ELECTRIFICATION IN BAYELSA STATE

ATLANTIC INTERNATIONAL UNIVERSITY

HONOLULU, HAWAII

2011

TABLE OF CONTENT

OVERVIEW ON THE PROBLEMS OF RURAL ELECTRIFICATION IN BAYELSA STATE

1.0   INTRODUCTION

Rural electrification is a global challenge especially in the developing countries. The problem of rural electrification is acute in Nigeria because the country has not been able to provide adequate supply to the connected consumers. The case of Bayelsa State is peculiar because the geographical position of the state offer various problems to rural electrification. The state is bounded by rivers, estuaries, creeks and stagnant swamps that pose problems to rural electrification in terms of settlements, accessible roads, good available means of transport etc. The climatic conditions and flooding are factors militating against the rapid growth of this area. Lack of industrial customers, low load forecast and very long distances from the grid with huge cost of installation, maintenance and operations are key factors.

Bayelsa State is dissected centrally by longitude 6° East and latitude 4.5° North. Bayelsa State with a population of 1,121,693 spread over a land area of 12,000 sq kilometer most of which is water or wet lands. Its definitive boarder is the 185 kilometer of coast line through which its many rivers issue into the Atlantic Ocean (Adekpoju et al 2007, ECN 2004 Sambo 2006). There are few big cities due to the flooding and erosion and some are almost cut off from the capital due to the area of settlement.

Bayelsa State is located in the heart of the Niger Delta and is one of the central areas that produce the resources (Oil and Gas) which account for about 97 percent of the National total earning (Sambo 2008, Igbinovia and Orukpe 2007). The operational base of the oil companies are not located within the source of the revenue, thereby making the place poor, rejected with no impact of such huge revenue taken out of this areas. These areas of Bayelsa State was regarded as rural areas before the creation of the state, therefore it has little or no development. The Oil Companies contributed to the level of poverty because the administrative headquarter of these companies are out of state and are far from the source of the oil deposit. The outcome of this was denial of the people from all social amenities, unemployment, rejection, poverty and devastation of the ecological sources due to spillages and gas flaring (Igbinovia Orukpe 2007).

Due to the outcry, the government is trying to put certain things in place for the development of the Niger Delta, therefore it is necessary to study the problems facing the environment so

that policies, designs and construction will be based on facts concerning the development of the area.

The socio-economic development processes revolve around suitable and sustainable power supply. In fact, it is the nucleus of operations and subsequently the engine of growth for all sector of the economy. It also determines the living standard of the people and as well stop the immigration to urban areas. It is therefore necessary to study the factors militating against the rural electrification project, using the existing one as a case study.

2.0   THE EXISTING SUPPLY

Prior to the existing Kolo gas turbine most of the Towns were supplied by the diesel generators through the rural electrification scheme by the Rivers State utilities board. These systems were inadequate and far from been efficient.

Betwe4en 1980 – 1983, the Kolo creek gas turbine was constructed and commissioned. This gas turbine was intended for rural supply and at that period served only three local government areas. The installed capacity was 40MW with two 11KV plants.

From the creation of Bayelsa State in 1996, due to resettlement in the capital (Yenagoa), the need for power supply increased, with the Kolo Creek gas turbine as the only source of supply (Idoniboye-Obu and Odubo 2010, Pabla 2006) .The power station operates on 33kv busbar at the Kolo Gas turbine. The distribution prior to the intervention of the National Electric Power Authority is in fig. 1.

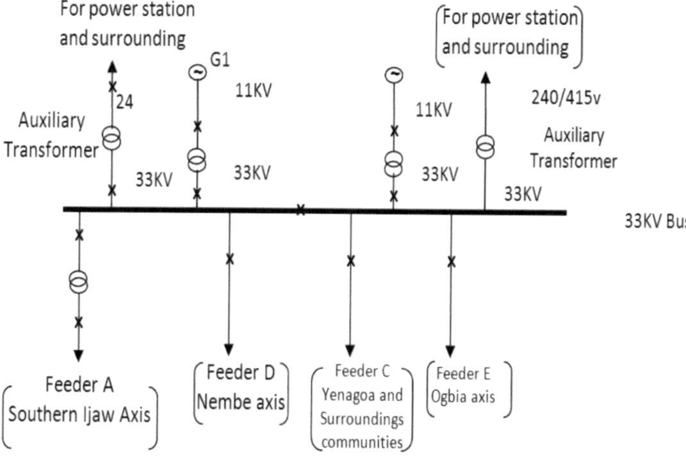

Fig. 1: Power system layout of Kolo Creek Gas Turbine

From the completion of the Gbarain Toru Gas turbine in the state about 20MW power was interconnected through the main Busbar of Kolo Gas turbine. The upgraded grid is shown in fig 2.

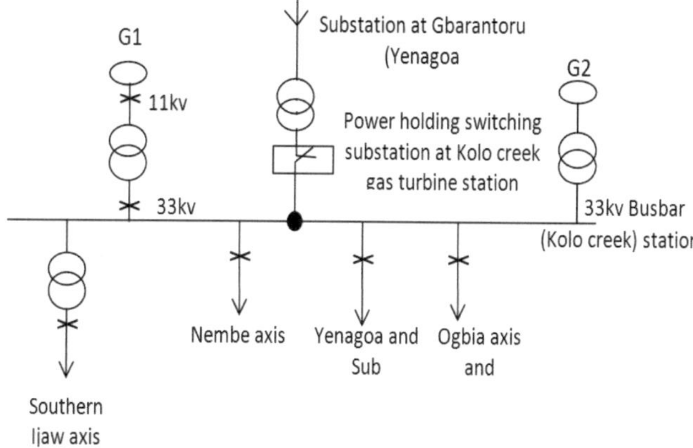

Fig. 2: Post grid connected power system

The main feeders are the ones serving the Southern Ijaw axis (Feeder A). From a 33/66kv transformer it is transmitted through 72km serving a main substation of 33kv after stepping down. Three sub-feeders go from the substation to different areas in the southern Ijaw local government. The feeder D which is designated as (Nembe Axis), serve most of the Ogbia town and presently stop at Akiplelai. The feeder C serves Yenagoa and some surrounding communities.

The power supply to Yenagoa through this feeder was grossly inadequate due to the resettlement as a result of state creation.

Presently, National Electric Power Authority supplements part of the supply in Yenagoa.

From the Gbarain Toru gas turbine a 33kv line run to Kaiama but due to lack of equipment upgrading the existing 11kv is used for transmission.

Table 1: The power distribution in Bayelsa State

S/N	Local Government Area	Installed Percentage %	State Utility	NEPA	Others
1.	Yenagoa LGA	60	Gas turbine		-
2.	Yenagoa (Capital)	80	Gas turbine	NEPA	-
3.	Ogbia LGA	85	Gas turbine	-	-
4.	Nembe LGA	No coverage	-	-	Generators
5.	Brass LGA	50	-	-	Agip gas turbine
6.	Ekenemo LGA	No coverage	-	-	Generators
7.	Sagbama LGA	No coverage	-	-	Generators
8.	Southern Ijaw	40	Gas turbine	-	-
9.	Kolokuma/Opokuma	20	-	NEPA	Generator

3.0 SUBSTATION FEEDER AND COST ANALYSIS

From fig 1 and 2, the voltage of feeder A is step up to 66kv by the transformer (33/66kv) and transported to Angiama substation at a busbar voltage of 33kv. The data for the feeder are given as:

Voltage level   =   66kv

No. of poles   =   250

Length of feeder   =   24km

Steel crossing tower =   1

Size of wire   = $150mm^2$

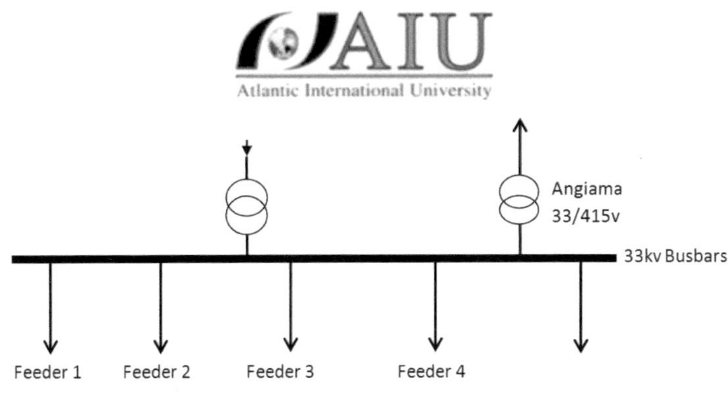

Fig. 3: Substation B

Table 2: Materials for Feeder 1 – 4

		Feeder 1	Feeder 2	Feeder 3	Feeder 4
No of Towns		6	12	7	14
No of transformer	PM	3	11	6	9
	GM	5	3	1	5
No of poles		350	320	242	580
Length of Feeders		35km	35km	25km	60km
Steel Tower Crossing		2	4	1	4
Total installed load KVA		975KVA	1.3MVA	650KVA	1.725MVA
Total Current Demand		1356.5A	1810AMP	904.3A	2399.8AMP

11B: Pm = Pole mounted transformer (50 – 100kva)

GM = Ground mounted transformer (200 – 300kva)

Table 3: Material for Feeder A, B, C, D and E

	Feeder A	Feeder B	Feeder C	Feeder D	Feeder E
No of Towns	39	16	13	6	14
No of transformer PM	29	6	5	4	9
GM	14	10	9	2	5
No of Poles	1492	520	400	250	580
Length of feeders	155km	54km	42km	250km	58km

Steel Tower Crossing	11	2	1	-	4
Total installed load	4.65MVA	1.95MVA	1.73MVA	600KVA	1.28MVA
Total Current Demand	6470.3A	2713Amp	2406.8A	835Amps	1774Amp

4.0 ANALYSIS OF FEEDERS / LOAD CONSUMPTION

In order to determine the economy of electrifying these areas it is necessary to consider the power demand by consumers in the area and cost of the power delivery. For the purpose of analysis Feeder 1 and Feeder D were used for both cost analysis and the available power.

Feeder 1:

From table 2, the data are as follows:

No of Town connected to the feeder = 6

No of transformers (Pm = 3 and GM = 5)

No of poles 33kv is 350 (NOs)

Length of feeder is 35km

Steel Crossing Tower 4

For transformer of 100KVA (Pole mounted) the current rating is

$$I = \frac{\rho}{\sqrt{3}\ V_2} = \frac{100000}{\sqrt{3}\ \times\ 415} = 138\ Amps$$

In a rural community, if it is assumed that an installation (house) consumes 10Amps, then for 100KVA transformer (138Amps), the number of consumer for maximum loading is 13 to 15 households. Also for a 200KVA transformer the maximum current loading is 278Amps, therefore about 27 to 30 customers may be supplied. For a feeder with eight (8) transformers (5 GM, 3 PM), the total expected customers will be 148 to 190 households.

The total cost for taken the power to these communities is given in table 3.

Table 4: Bill of Quantity for Feeder 1

S/N	Material Description	Number	Unit (USD) Cost	Total Cost
1.	Transformer – Pole mounted – ground mounted	3 (100KV) 5 (200KVA)	600 10,000	18000 50,000
2.	33kv concrete poles	400N	150	60,000
3.	33kv cross arms (steel)	400	35	14,000
4.	Lightning Arrester	10 units	670	6700
5.	Disc insulators	2000	20	40,000
6.	Pin insulator	900	16	14,400
7.	J-hook	200	10	2000
8.	Screws bolts and nuts	10+	12,000	12,000
9.	Steel Crossing Tower	4 pairs	14,000	105,000
10.	Length of wire	150km	3/m	450,000
11.	Stay wire	2.4km	2/m	4800
12.	Earth rod	350	3.5	1225
13.	Stay rod	30	6	180
14.	Stay Insulator and blocks	10t	14000	14000
15.	Channel Iron	20	6	120
				793425 USD

SUMMARY

Total Material Cost	=	793425
Transport	=	238027.5
Labour Cost	=	317370
Site Accommodation/ Security etc	=	155,000
Community	=	1568.5
	=	**1505391 USD**

For feeder D, the data for the feeders are as follows:

No of Town (Village)	=	6
No of Transformer PM	=	4
No of Transformer GM	=	2
No of Poles 33KV	=	250
Length of feeders	=	25KM
Length of Wire	=	97500M
The total KVA	=	600KVA

The total expected numbers of customers are 120, that is, each of them taken 10Amps or less.

The summary of the total cost is shown in table 5

Total Material Cost	=	832561
Transport	=	231245
Labour Cost	=	350178
Site Accommodation	=	160000
Community	=	2500
	=	**USD 1576484**

4.1   Calculation of Voltage Drop for Feeder

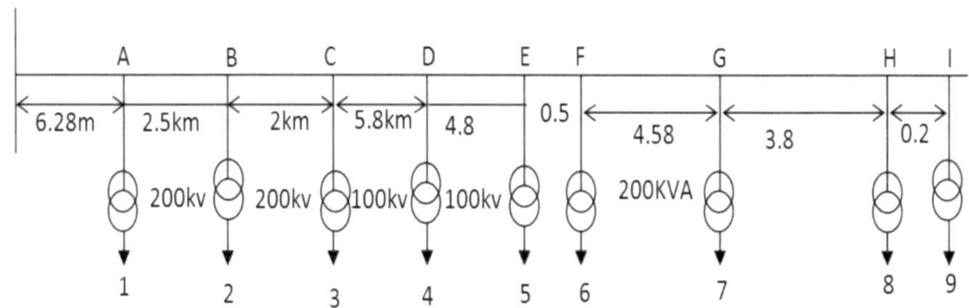

Fig. 4: Bill of Quantity for Feeder 1

Total transformer capacity shown in fig.

$$6 \times 200KVA = 1200KVA$$

$$3 \times \frac{100KVA = 300KVA}{Total \quad 1500KVA}$$

Let the diversity factor DF = 2.5 and load power factor to be 0.8.

The maximum demand load $= \dfrac{1500}{2.5} = 600KVA$

Maximum feeder current $= \dfrac{600}{\sqrt{3} \times 33} = 10.5 Amps$

OA	=	1500 x 6.2	=	9000
AB	=	1300 x 2.5	=	3250
BC	=	1100 x 2.0	=	2200
CD	=	1000 x 5.8	=	5800

DE	=	900 x 4.8	=	4320
EF	=	700 x 0.5	=	350
FG	=	500 x 4.58	=	2290
GH	=	400 x 3.8	=	1320
HI	=	200 x 0.2	=	40
Total KVA x KM			=	28770

Percentage Voltage drop $= \dfrac{Voltage\ drop\ per\ KM\ KVA \times total\ KM\ KVA}{DF}$

For an ACSR conductor the KM-MVA, for voltage regulation at 0.8pf for 150mm^2 conductor is

$$\frac{30/141.95 \times 28770}{1000 \times 2.5} = \frac{172620}{354875} = 2.43\%$$

5.0   DISCUSSION

From analysis on table 2, the lines covered long distances for few sparingly populated villages. Considering the number of transformer for each village it shows that the serviceable houses were few. From feeder 1 and feeder D it was analyzed that each pole mounted transformer can serve only 12-15, each customer consuming about 10Amp and the ground mounted transformer could serviced up to 24 to 30 customer taking the same load.

From feeder 1, the cost of sending power to only six villages cost about 1.6 million U.S dollars excluding the internal distribution. It could be seen that the rural electrification in Bayelsa is capital intensive.

If it is assumed that in every month by the load demand each consumer is to pay twenty U.S dollars, since there are 195 customers, the total payment for the place under consideration (feeder 1) is 3,900 U.S dollars for a month. This amount is only able to pay ten (10) average workers. Therefore, it is grossly inadequate for office maintenance and salary payment. It could be seen that money recovered from consumer cannot be sufficient for transport and maintenance of equipment (Togola 2005, Obsen 2004,Barr. And Thomas 2005).

NATURE OF RURAL LOADS

In an electric power system different loads on the system can be identified. These could be domestic, commercial industrial and municipal loads. In the rural areas, especially the areas under consideration, the loads are mostly domestic except in some fairly big cities were commercial loads could be installed (Fagbenle et al 2008, Suhartone 2009). The load demand of the rural area for a day could be presented as in fig.5

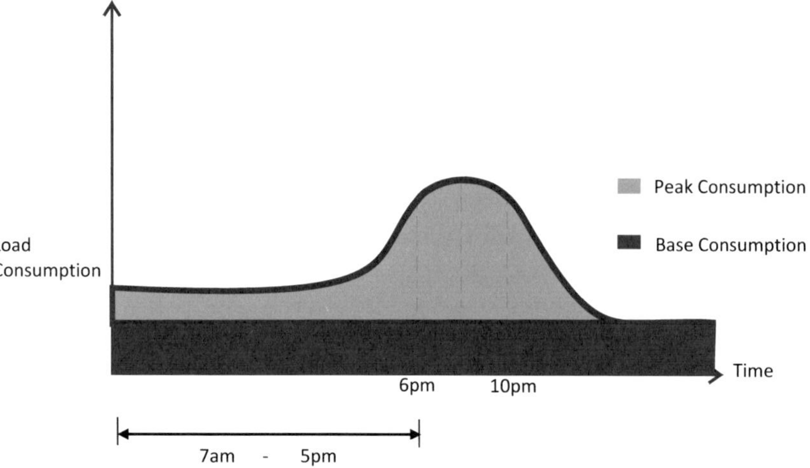

Fig. 5: Electricity Consumption Pattern for the day

The peak period in the year is always the end of the year (December and early January). Other period may be the festival periods when people visit home. In the day the peak period is the evening periods (6pm to 9pm).

Fig. 6: Electricity Consumption Pattern for the Year

In an electric power system, since there are no commercial and industrial loads, low tariff characterize these areas.

6.0   RECOMMENDATIONS

The grid system is far from these areas therefore bringing power to these areas are extremely difficult and is high capital intensive therefore the following recommendations are made:

❖ The area should be studied and other forms of energy sources be used, that will reduce government subsidy and enhance better development.

❖ The renewable energy sources, especially solar energy generating sources are available and could be used in remote areas. Villages could be grouped together and a common source of energy could be used to reduce transmission cost. Other sources could be mini hydro- turbine, wind etc.

❖ In the coastal areas, the wind energy source is the best option for such an area. Even if HVDC are used the cost of installation of the rectifier – converter substation will outweigh the cost of transmitting power.

❖ The grid system could be used for bigger towns and those areas where commercial and industrial users could pay for the running cost.

7.0   CONCLUSION

The low population density, low consumption and difficult terrain are causes of the problems of rural electrification scheme in Bayelsa. Therefore the state suffers from such essential amenity.

The oil companies operate only the flow stations therefore the presence of these companies is not felt. As a result of these no small or medium scale industries are attracted, therefore the provision of electricity is not encouraged. This has resulted to urban migration, abandonment of the area even by the educated indigenes. Due to low consumption and the high cost of maintenance the government has abandon the existing supply and provision of new lines are not feasible.

Long distances from the grid system results to greater electricity losses that require more expensive customer support and equipment maintenance but the dwellers are poor with very low income. It could be seen that even if a tariff system is introduced, the rural electrification scheme will need a high subsidized government support or some agency to properly deliver the required services. Therefore, the scheme is not attractive and cannot attract competitors. As was said earlier the present scheme covers only three local governments and has been supported by the state government but due to political reasons and sometimes cash-constraint government, make the availability of electricity a dream that is never realized.

Due to the terrain, the transportation system, the low consumption (no customer support) and the tariff system, the electricity scheme is as low as ten (10) percent in the state. Off-grid services based on renewable energy if properly studied will improve the availability of electricity.

REFERENCES

[1] Adepoju, G.A., Ogunjuyigbe, S.O.A., Alawode, K.O., 2007: The Pacific Journal of Science and Technology. 8(1): 68 – 73.

[2] Alfareo, H.K., Nazeeruddin, M., 2002: Electric Load Forecasting Literature survey and classification of methods: International Journal of Systems Sciences 33(1): 23 – 34.

[3] Balderer, C. 2007. The Network reliability problem with applications in electricity networks (Master Thesis) ETU Zurich – Department of Mathematics, Institute of Operation Research.

[4] Barr., S., Thomas, JP., 2005: The role of renewable energy in the development of productive activities in rural West Africa: the case of Senegal.

[5] ECN, 2004. National Energy Policy, Federal Republic of Nigeria. Published by the Energy Commission (ECN).

[6] Fagbenle, R.Q., Adeleja A.Q., Bellow, A.K., 2008: Modeling of Wired Energy Potential in Nigeria. Final Report of the Power sector Reform Committee, Abuja.

[7] Idoniboye-Obu, D.C, Odubo, G.F., 2010: Electrical Load Forecast in Bayelsa State of Nigeria by the year 2020, The Nigerian Journal of Industrial and Systems Studies. 9 (1): 22-27.

[8] Igbinovia, S.O., Orukpe, P.E., 2007: Rural Electrification: the propelling force for rural development of Edo State, Nigeria.

[9] Komolafe, O.A., Omoigui, M.O., 2006: An Assessment of Reliability of Electricity Supply in Nigeria: 4th International Conference on Power Systems Operations and Planning Accra – Ghana.

[10] Olesen, G.B., 2004. Inforse Vision: 100% renewable in 2050, Sustainable Energy News, 45.

[11] Pabla, A.S., 2006. Electric Power Distribution, Tata McGraw – Hill Publishing Company Limited, New Delhi, pp. 147 – 222.

[12] Sambo, A.S., 2006. Renewable Energy Electricity in Nigeria the way forward: Renewable Electricity Policy Conference. Pp. 1 – 42.

[13] Sambo, A.S., 2008. Matching Electricity Supply with Demand in Nigeria: International Association for Energy Economics: pp. 321 – 336.

[14] Sambo, A.S., 2008. The Role of Energy in Achieving Millennium Development Goals (MDGs) National Engineering Tech. Conference (NETEC): 1 – 42.

[15] Suhartono, Endharta, A.J., 2009. Short Time Electricity Load Demand Forecasting in Idonesia by using Double Seasonal Recurrent Neural Network: International Journal of Mathematical Models and Methods in Applied Sciences 3 (3): 171 – 178.

[16] Togola, I., 2005. Renewable energy solution perspectives for Africa Mali – Folke Centre.